WALTER HAYS SCHOOL LIBRARY

S0-EFL-262

DATE DUE

FEB 20 '89			
Parent Bull			
OCT 10 1990			
NOV 03 1992			
T-16			
T. 16			
T-1			
T-2			
T-1			
JAN 19			

```
523.3      Langley, Andrew
L               The moon.   New York, Watts, 1987.
                32 p.  ill. draws. photos. (part col.)
c.1             Glossary, p.31
```

WALTER HAYS SCHOOL LIBRARY
PALO ALTO PUBLIC SCHOOLS

1. Moon   I. Title

EASY-READ FACT BOOKS

The Moon

Andrew Langley

Franklin Watts
London · New York · Sydney · Toronto

© 1987 Franklin Watts Ltd
Franklin Watts
12a Golden Square
London W1

First published in the USA by
Franklin Watts Inc.
387 Park Avenue South
New York, N.Y. 10016

Franklin Watts Australia
14 Mars Road
Lane Cove
NSW 2066

Phototypeset by Keyspools
Limited
Printed in Hong Kong

UK ISBN: 0 86313 575 7

US ISBN 0-531-10447-8
Library of Congress Catalog
Card No: 87-61349

Photographs:
NASA
Space Frontiers Ltd

Illustrations:
Christopher Forsey
Product Support Graphics
Michael Roffe

Design:
Janet King
David Jefferis

Technical consultants:
Nigel Henbest MSc, FRAS
Iain Nicholson BSc, FRAS

Note: The majority of illustrations in this book originally appeared in "The Moon": An Easy Read Fact Book

Contents

Our neighbor in Space — 4
The Earth and Moon compared — 6
Phases of the Moon — 8
The Moon in close-up — 10
Lunar craters — 12
Seas and oceans — 14
Mountains and valleys — 16
Man on the Moon — 18
Lunar explorers — 20
Future Moonbase — 22
Eclipses — 24
Moon Atlas 1 — 26
Moon Atlas 2 — 28
Moon facts — 30
Glossary — 31
Index — 32

Our neighbor in Space

The Moon is the Earth's nearest neighbor. It goes around the Earth at a distance of 239,000 miles (384,000 km).

The Earth and Moon compared

The Earth and Moon are very different. The Moon is much smaller than the Earth. The Moon has no life as far as we know. There is no water or air on the Moon.

The Moon is 2,160 miles (3,476 km) across, about one quarter the size of the Earth.

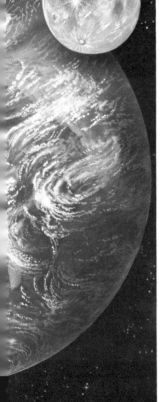

Footprints on the Moon will never change, as there is no wind.

Phases of the Moon

The Moon shines because it reflects light from the Sun. As it goes around the Earth we see different amounts of its sunlit side. The same side always faces us. We never see the far side.

Quarter Moon

Half Moon

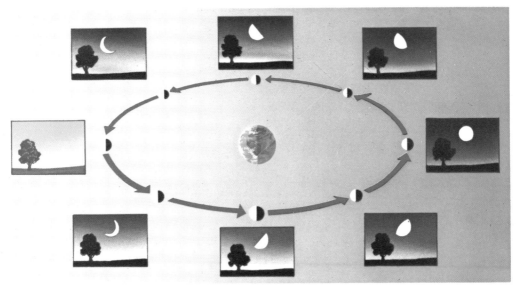

The Moon circles the Earth once every $27\frac{1}{3}$ days. There are eight phases of the Moon each month.

Full Moon

Three-Quarter Moon

The Moon in close-up

For hundreds of years people studied the Moon through telescopes. Today we can see it more closely. Space probes have taken pictures of the Moon's surface – even its far side.

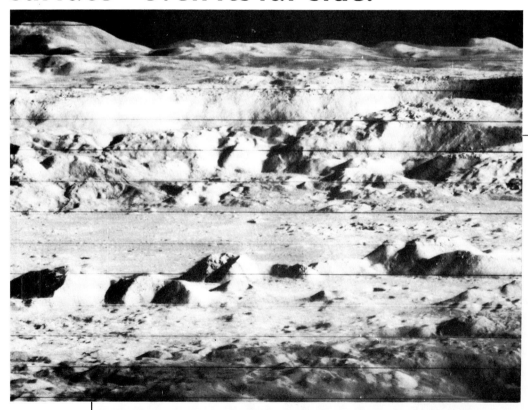

The Lunar Orbiter Space probe

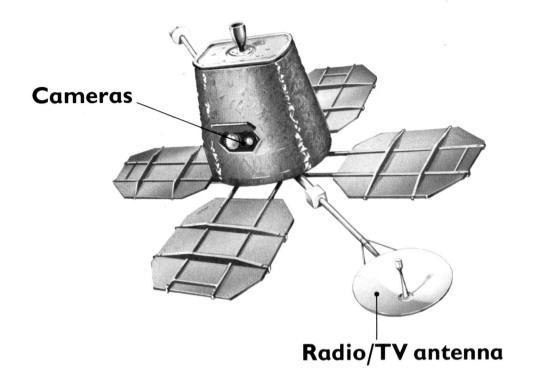

Cameras

Radio/TV antenna

The crater Copernicus seen through a telescope from Earth.

Lunar craters

The Moon is pitted with craters. Most are just tiny dents. But some are vast basins hundreds of miles wide.

0 155 miles (250 km)

The size of Italy compared with the crater Clarius on the Moon.

The craters were made by a shower of rocks called meteorites. Many meteorites were no bigger than grains of sand, but some were huge. This picture shows a big meteorite hitting the Moon. The surface of the Moon is now covered with dust, boulders and rock chips.

Seas and oceans

The dark areas on the Moon are vast deserts. People once thought they were seas and oceans. They were made when giant meteorites hit the Moon and cracked the surface. Molten rock bubbled up from below and grew hard.

The arrow shows the Ocean of Storms where the astronauts, seen right, landed in 1969.

Mountains and valleys

The Moon has many mountains and valleys. The highest mountains are almost as tall as Mount Everest on earth. The deepest valley is nearly 1,312 ft (400 m) deep. The far side of the Moon has more craters and mountains than the near side.

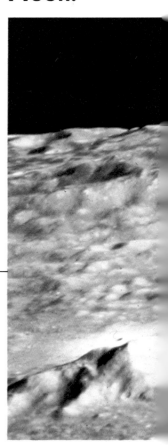

A view of the far side of the Moon.

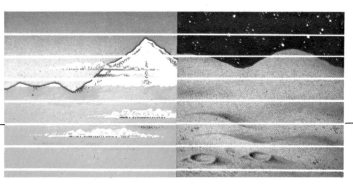

The far side has thousands of craters.

Mount Everest compared with some peaks on the Moon.

Man on the Moon

Neil Armstrong became the first man to step on the Moon, in 1969. He and a crewmate spent one day on the Moon. They landed in the Lunar Module seen here. There have been five Moon landings since then.

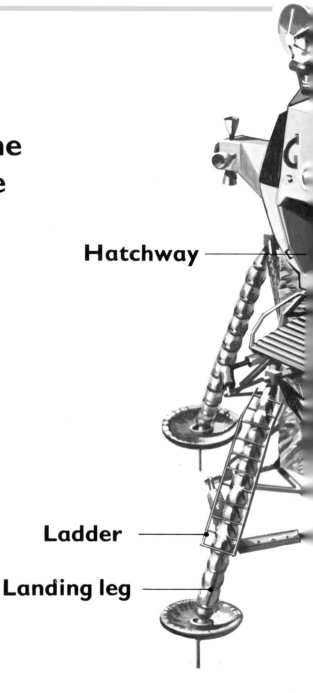

Hatchway

Ladder

Landing leg

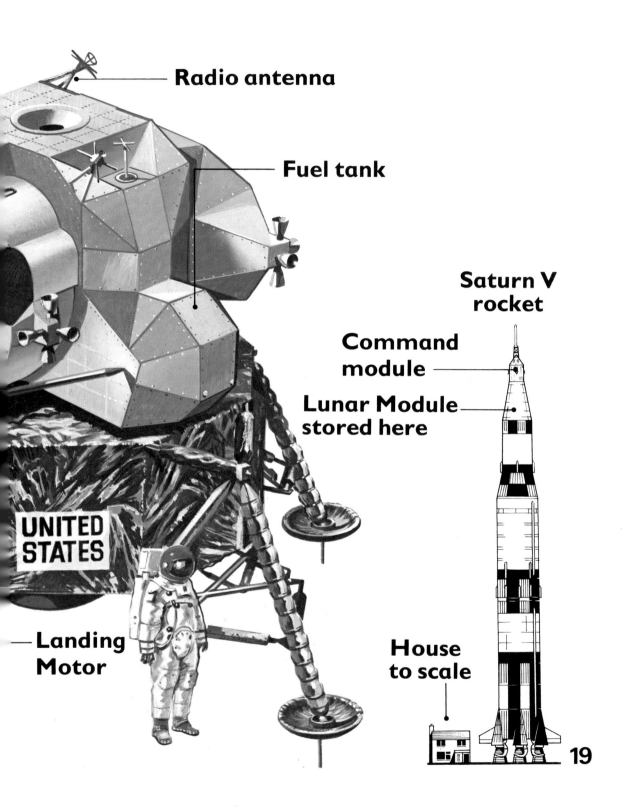

Lunar explorers

The astronauts had many jobs to do. They traveled over the Moon in a "Lunar Rover." They collected rocks to take back to Earth, and set up pieces of equipment.

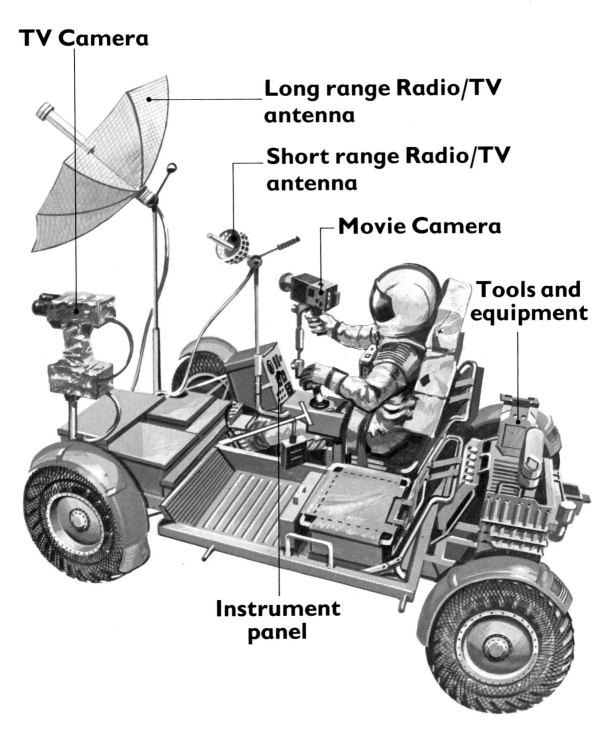

Future Moonbase

Will there by any more trips to the Moon? New types of spaceship should make traveling easier and cheaper. It may be possible to build a Moonbase like the one shown here. In thirty years' time many people might be living and working on the Moon.

Eclipses

The Moon sometimes passes between the Earth and the Sun, blocking out the Sun's light. This is called a *total eclipse* of the Sun. The Moon is eclipsed when the Sun's light is blocked by the Earth.

How an eclipse of the Moon happens

How an eclipse of the Sun happens

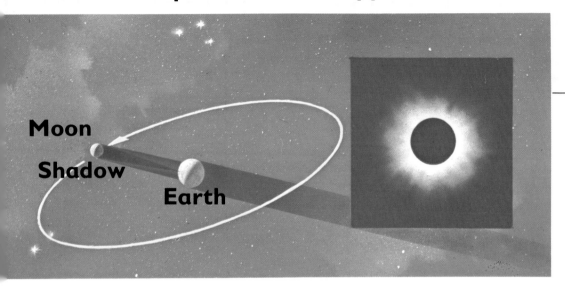

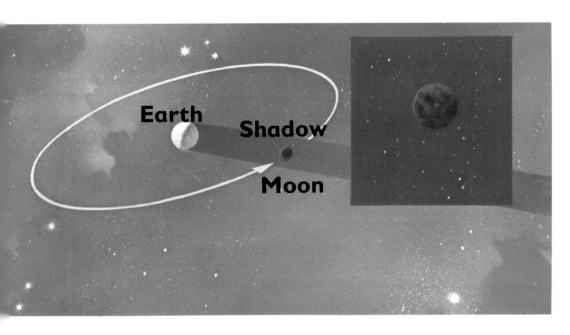

Moon Atlas 1

This map shows the upper half of the Moon as we see it from Earth.

0   155 miles (250 km)

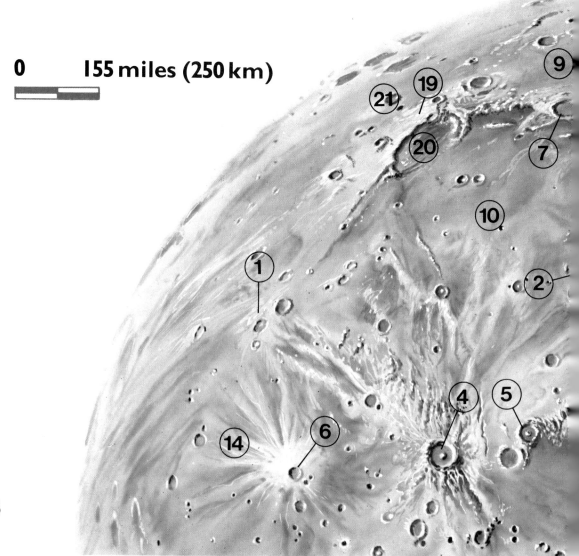

Craters
1 Aristarchus
2 Archimedes
3 Aristoteles
4 Copernicus
5 Eratosthenes
6 Kepler
7 Plato

Maria ('seas')
8 Mare Crisium
  Sea of Crises
9 Mare Frigoris
  Sea of Cold
10 Mare Imbrium
  Sea of Rains
11 Mare Serenitatis
  Sea of Serenity
12 Mare Tranquillitatis
  Sea of Tranquility
13 Mare Vaporum
  Sea of Vapours

Oceanus ('ocean')
14 Oceanus Procellarum
  Ocean of Storms

Mountains
15 Apennines
16 Alps
17 Caucasus
18 Haemus
19 Jura

Sini ('bays')
20 Sinus Iridum
  Bay of Rainbows
21 Sinus Roris
  Bay of Dews

Landing sites
22 Apollo 11
23 Apollo 15
24 Apollo 17

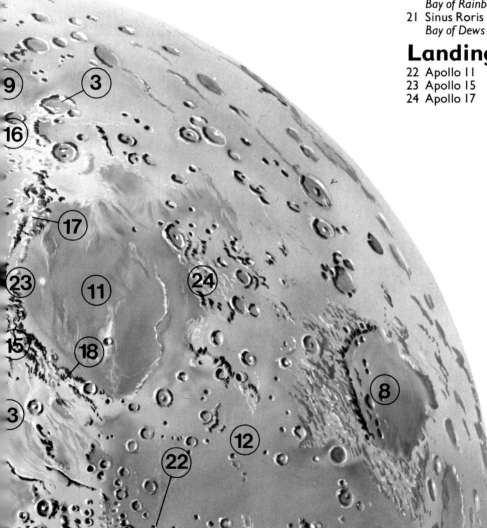

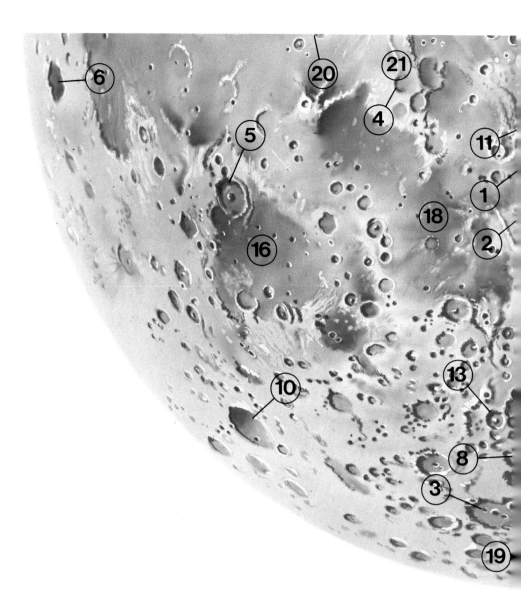

Moon Atlas 2

The bottom half of the Moon has many large craters.

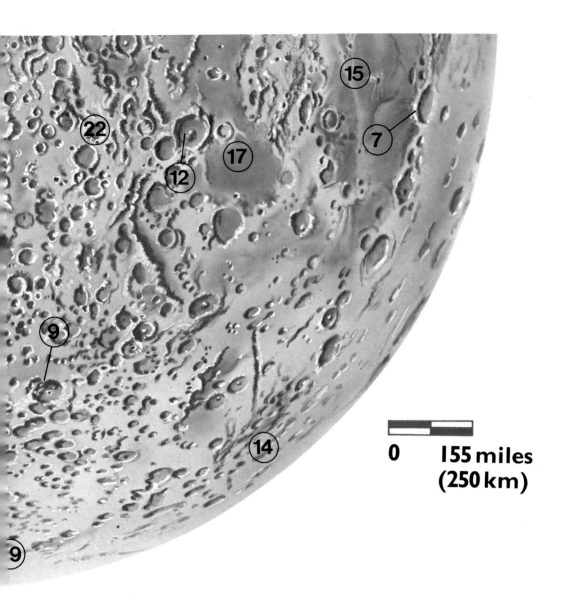

Craters
1 Alphonsus
2 Arzachel
3 Clavius
4 Fra Mauro
5 Gassendi
6 Grimaldi
7 Langrenus
8 Maginus
9 Maurolycus
10 Schickard
11 Ptolemaeus
12 Theophilus
13 Tycho

Maria ('seas')
14 Mare Australe
   Southern Sea
15 Mare Fecunditatis
   Sea of Fertility
16 Mare Humorum
   Sea of Humours
17 Mare Nectaris
   Sea of Nectar
18 Mare Nubium
   Sea of Clouds

Mountains
19 Leibnitz

Landing sites
20 Apollo 12
21 Apollo 14
22 Apollo 16

Moon facts

The Moon is about 239,000 miles (384,000 km) away from Earth.

By studying Moon rocks, scientists have worked out that the Moon is about 4.6 billion years old.

The Moon turns very slowly on its axis. One "day" on the Moon lasts almost as long as a month on Earth.

In 1959 the Soviet space probe Luna 1 was the first to pass beyond the Moon. Later that year, Luna 3 took the first photographs of the Moon's far side.

The Moon orbits the Earth once every $27\frac{1}{3}$ days.

The pull of gravity is much weaker on the Moon than it is on Earth. When astronauts tried to walk, they found that they bounced into the air!

There have been six successful manned landings on the Moon. The last crew spent a record 75 hours on the Moon's surface in 1972.

To their surprise the astronauts discovered that there is glass on the Moon. Small beads of it were long ago formed by great heat.

Glossary

There are some technical words used in this book. This is what they mean.

Astronaut
One of the crew of a space ship.

Crater
Circular dent on the Moon's surface caused by the impact of a meteorite.

Eclipse
The moment when the Moon blocks the Sun's light from the Earth or the Earth throws a complete shadow over the Moon.

Far side
The side of the Moon we cannot see from the Earth. It is also called the "dark side" of the Moon.

Gravity
The force of attraction which pulls together everything in the Universe. Bigger objects, like the Earth, pull more strongly than small ones, like the Moon.

Meteorite
A chunk of space rock. Both Earth and the Moon are still sprayed by showers of meteorites every day.

Near side
The side of the Moon seen from the Earth.

Planet
An object which goes around a star, such as the Sun, and is lit by its light.

Index

Armstrong, Neil 18
Apollo 31
Astronaut 20, 30, 31

Crater 12, 13, 16, 28, 31

Desert 14

Earth 4, 6, 8, 9, 16, 24, 26, 30
Eclipse 24, 31

Far side 8, 31

Glass 30
Gravity 30, 31

Luna 1 30
Lunar Rover 20
Lunar module 18

Meteorite 13, 14, 31
Molten rock 14
Moonbase 22
Mountain 16

Phases of the Moon 8

Sea 14
Solar panel 31
Space probe 10
Space ship 22, 31
Sun 8, 24, 31